W9-AUY-245

THE CHANGING FACE OF
MALAYSIA

Text by AIDEN GLENDINNING and JIM HOLMES
Photographs by JIM HOLMES

Raintree
Chicago, Illinois

Published by Raintree, a division of Reed Elsevier, Inc.
Chicago, Illinois
Customer Service 888-363-4266
Visit our website at www.raintreelibrary.com

For information address the publisher:
Raintree
100 N. LaSalle
Suite 1200
Chicago, IL 60602

Library of Congress Cataloging-in-Publication Data:

Glendinning, Aiden.
  Malaysia / Aiden Glendinning and Jim Holmes.
      p. cm. -- (Changing face of--)
Summary: Presents the natural environment and resources, people and
culture, and business and economy of Malaysia, focusing on development
and change in recent years.
Includes bibliographical references and index.
  ISBN 0-7398-6830-6 (lib.bdg.)
  1. Malaysia--Juvenile literature. [1. Malaysia.] I. Holmes, Jimmy.
II. Title. III. Series.
  DS592.G575 2003
  959.5--dc21

                              2003009742

Printed in China, bound in the United States.

08 07 06 05 04
10 9 8 7 6 5 4 3 2 1

Acknowledgments
The publisher would like to thank
the following for their contributions
to this book: Rob Bowden—statistics
research; Peter Bull—map
illustration; Nick Hawken—statistics
panel illustrations. All photographs
are by Jim Holmes.

Contents

Penang—Pearl of the Orient

Since the 1500s the island of Penang has been a gateway to business in Southeast Asia. By 1786 the East India Company, whose business expansion laid the foundations of the British Empire, had made Penang a major trading post. George Town, the main port, soon became a bustling commerce center, attracting traders from all over the east. Visitors fell in love with Penang's beauty, and the island became known as the "Pearl of the Orient."

▲ *Locals in George Town, Penang often meet up for a chat at tea stalls.*

Penang's importance in world trade declined as nearby Singapore's grew bigger. After Malaysia gained independence from Britain in 1957, business boomed again. Tourism became a major source of income for the island. The government also saw it as the perfect place to develop new high-tech industries, and the number of computer-chip factories and semiconductor plants now on Penang has led some people to call it "Silicon Island."

The recent rapid development of industry and tourism has caused a lot of concern in Penang. Fishermen complain that marine life is badly affected, and the number of tourists is falling as the island's beauty is becoming spoiled. Today, the "Pearl of the Orient" faces some big challenges.

▶ *Tourists enjoy a relaxing ride in an old trishaw through the narrow streets of George Town, Penang.*

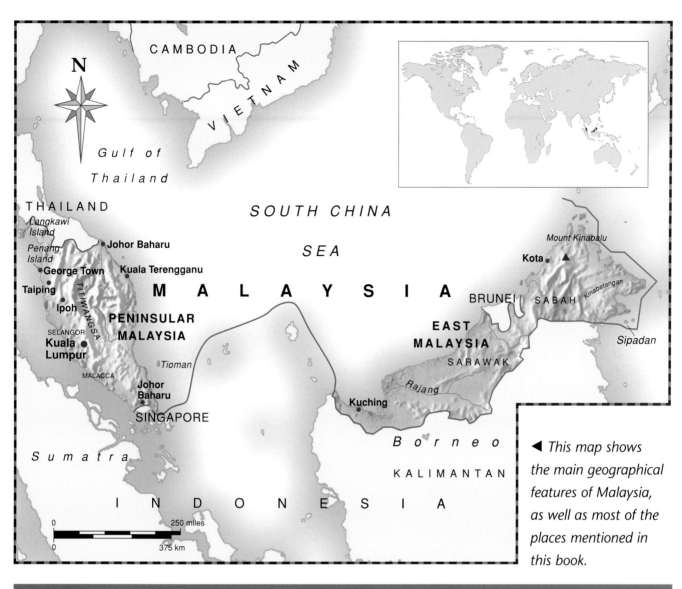

This map shows the main geographical features of Malaysia, as well as most of the places mentioned in this book.

Places labeled on the map:

CAMBODIA
VIETNAM
Gulf of Thailand
THAILAND
Langkawi Island
Penang Island
Johor Baharu
George Town
Kuala Terengganu
Taiping
Ipoh
TITIWANGSA
PENINSULAR MALAYSIA
SELANGOR
Kuala Lumpur
MALACCA
Tioman
Johor Baharu
SINGAPORE
Sumatra
INDONESIA
SOUTH CHINA SEA
MALAYSIA
Mount Kinabalu
Kota
SABAH
Kinabatangan
BRUNEI
EAST MALAYSIA
SARAWAK
Sipadan
Rajang
Kuching
Borneo
KALIMANTAN
250 miles
375 km

MALAYSIA: KEY FACTS

Area: 177,355 square miles (329,847 square kilometers)

Population: 24.01 million (mid-2002 estimate)

Population density: 28 people per square mile (73 people per square kilometer)

Capital city: Kuala Lumpur (1.3 million)

Other main cities: Ipoh (0.53 million), Petaling Jaya (0.43 million), Johor Baharu (0.40 million), Kota Kinabula (0.35 million)

Highest mountain: Mount Kinabalu 13,455 feet (4,101 meters)

Longest river: Rajang 352 miles (567 kilometer); Kinabatangan 348 miles (560 kilometers)

Main language: Bahasa Malayu (Malay)

Major religions: Islam (52 percent), Buddhism (17 percent), Taoism (12 percent), Christianity (8 percent), Hinduism (8 percent), Animism (3 percent)

Money: Malaysian ringgit (1 ringgit = 100 sen, 3.78 ringgit = $1)

2 Past Times

About 4,500 years ago, groups of people whom historians call the Proto Malays moved to Southeast Asia from China, joining the aboriginal peoples already there. Migration waves from India, China, Siam (now Thailand), and Arab lands in the Middle East mixed with the Proto Malays to create a loose grouping known as the Deutero Malays. The Deutero Malays mixed with Indonesian groups to form the Malays.

Early Indian and Greek writers called the Malay peninsula a "Land of Gold" because of its rich trade. Malay rulers adopted many customs from Indian and Chinese traders, including the Hindu religion, and later on, Buddhism. Islam was introduced in the 15th century as a result of contact with Muslim merchants.

▲ *Most Malay females wear Islamic dress when out in public.*

European powers established colonies in Malay ports to control the rich trade in gold, wood, spices, and tin. In the 19th century, Malaya came under British control. During World War II (1939–1945), Malay was captured by the Japanese.

◄ *Modern Kuala Lumpur is a city with an exciting mix of architecture. This is the Jamek mosque in downtown Kuala Lumpur.*

Malaya to Malaysia

Malaya won independence from Great Britain in 1957, and Tunku Abdul Rahman became the country's first Prime Minister. In 1961 Singapore, Sabah, and Sarawak joined Malaya in a federal union called Malaysia. Singapore left the union in 1965. The country struggled through a three-year confrontation with Indonesia, and later with clashes between its Malay and Chinese communities, before establishing a stable state. Since the 1980s Malaysia has experienced tremendous growth and prosperity, with new industries creating a wealthier and more peaceful society.

▶ *The Petronas twin towers, headquarters of the state oil company, are the tallest buildings in the world and a symbol of Malaysia's economic progress.*

IN THEIR OWN WORDS

My name is Mohammed Amer bin Imran, and I live in Kota Baharu. I am sixteen years old. My grandad always tells me how hard life was when he was growing up and what it was like being a prisoner of the Japanese during World War II. I want to be a doctor when I get older, just like my mother and father. I want to help people if I can. My grandparents did not get much chance at higher education, so I feel lucky that I can study as much as I want. Kota Baharu is much bigger now than it used to be. It's more modern and convenient with more hospitals, roads, and public transportation. Who knows what Malaysia might achieve in the future—maybe we will even send somebody into space!

3

Landscape and Climate

Malaysia is separated into two regions, divided by the South China Sea: Peninsular Malaysia, also called West Malaysia, and East Malaysia. Thirteen *negeri*, or states, and three federal territories make up the Federation of Malaysia. Together with its islands, the country is almost the same size as Norway or New Mexico.

Peninsular Malaysia is part of mainland Southeast Asia. It is the more developed part of the country where most of the major cities are located. It is bordered by Thailand to the north, with the Andaman Sea along its west coast and the South China Sea to the east. Singapore lies at its southern tip, connected by a bridge.

▼ *The forests of Malaysia are often covered in mist.*

East Malaysia's states of Sabah and Sarawak cover the northern half of Borneo. Borneo is a huge island on which Brunei and the Indonesian province of Kalimantan are also located. The people of East Malaysia are mostly from ethnic groups that are quite different from the Malays of the West.

Mountains and coastal plains

The southern part of Peninsular Malaysia is relatively flat, but the north contains several mountain ranges, including the Cameron Highlands, whose cool forests are popular with visitors. At 13,455 feet (4,101 meters), Mount Kinabalu in Sabah is the highest peak in Southeast Asia.

Along the coasts of the peninsula are ancient mangrove forests full of rare plants. There is a wide, fertile plain on the west coast and a narrow coastal plain on the east. There is so much forest and mountain terrain that there is not much land for farming. As the population grows, natural features such as the rain forests and rivers are threatened.

▼ *This mangrove forest is a national park. The walkways allow visitors to explore the rich, but wet, environment.*

IN THEIR OWN WORDS

I am Roslan (on the left) and I am a farmer from Penang Island. I suppose any work is hard really, but farming seems to be one of the hardest jobs. Times have really changed in Malaysia, and there has been a big shift from agriculture to technology and other industries. We are building an underground water tank here. There used to be a stream that ran past the edge of the field, and we could grow crops all year round. Then, a few months ago, the stream just dried up. Without a steady flow of water, how can the farm survive? If you set your mind to it, though, you can do anything. As the environment changes, we have to change, too, and develop.

Ancient forests

The tropical rain forests that cover 61 percent of Malaysia are thought to be the oldest in the world. They were too far away from the ice sheets of the last Ice Age to be affected by the low temperatures, meaning that they are about 130 million years old. These forests cover an area almost as big as the whole state of Oregon.

▲ *The forests in the Titiwangsa mountain range near Fraser's Hill are among Malaysia's greatest treasures.*

Plentiful water

The country's two longest rivers, the Rajang (352 miles/567 kilometers) and the Kinabatangan (348 miles/560 kilometers), are in East Malaysia. Many of the steep mountain slopes of Sarawak and Sabah receive over 197 inches (5,000 millimeters) of rain per year, and fast-flowing rivers rush down the mountains. These can be dammed to provide hydroelectric power. Most of Malaysia's lakes were created by dam projects.

Sticky weather

The tropical climate is hot and humid all year. Two monsoons sweep the country from its coastlines throughout the year, drenching the plains and Malaysia's many mountains. There is no cold season, and average daily temperatures are between 68 °F and 86 °F (20 °C and 30 °C). There is so much water in the air that humidity is usually around 90 percent, making daily life rather sticky. It is this constantly humid climate that produced and maintains Malaysia's famous rain forests, which teem with animal and plant life.

▶ *Malaysia's tropical rainstorms can be very heavy.*

IN THEIR OWN WORDS

My name is Dr. Adzmi Yacoub and I'm a soil scientist from Taman Negra, near Kuala Lumpur. Soil is so important to us—basically it's what plants survive on. Soil is a very thin layer on top of the earth's rocks, and the deeper the soil, the better it should be. Soil is very fragile: When you take out trees—for logging, for example—you have to be careful not to disturb the topsoil. Roots hold the soil in place even on steep slopes, and if you are not careful you can get terrible landslides that cause a lot of damage after the trees have been chopped down. Heavy rainfall makes the problem even worse, and we can get up to 157 inches (4,000 milimeters) of rain in a four-month period.

4

Natural Resources

Powering the nation

Although oil was first discovered in the country in 1882, it was not until the 1970s that Malaysia began drilling in the ocean and producing large quantities of petroleum and natural gas. After a world oil crisis in 1973, the government took control of these valuable resources and created the Petronas state company to manage them. Crude oil is refined into gasoline and other products, mainly at big terminals in Sarawak. The country currently exports petroleum, but its oil reserves are not huge, and Malaysian people use more gasoline every year. By 2008 the country could be using more oil than it can produce. The oil wells could be dry by 2020, and scientists are trying to find ways to switch to solar power and biomass fuel made from palm oil. While over 90 percent of electricity is still generated from nonrenewable fossil fuels such as oil and coal, use of hydroelectricity is increasing, and there are plans to create many more dams.

▼ *Drivers fill up with LNG, liquid natural gas, which is much cheaper than gasoline and is a byproduct of oil exploration.*

Wealth from the earth

Tin has been exported from the Malay peninsula for centuries. Kuala Lumpur, the capital, was founded in 1857 by 87 Chinese tin miners in the middle of huge deposits of the metal. Products made from pewter (a mixture of tin and lead), especially from Selangor province, are still a valuable export, but tin mining has declined in Malaysia. Rubber trees, which produce the resin that is made into rubber, used to be a very valuable natural resource, but that industry is also now in decline. Timber remains an important industry, and a lot of wood is still cut from the forests of East Malaysia for sale.

▲ *Although natural rubber from trees has been replaced by synthetic products in many industries, big rubber plantations still exist.*

IN THEIR OWN WORDS

I am Mohammed, a taxi driver from Kuala Lumpur, and I am 29 years old. I drive a car powered by natural gas that I can get from a Petronas service station. Petronas is a very big company and their service stations are everywhere. It's also a Malaysian company, so it helps the country if we buy from them. Most taxis run on natural gas because it is half the price of gasoline. People tell me it is also cleaner and doesn't make so much pollution, which must be a good thing.

Plantations

Palm oil is made from the fruit and kernels of the oil palm tree. Malaysia is the largest producer of palm oil in the world. Ninety percent of it is used to make cooking oil and food products, but the government is promoting ways of converting the oil to other uses such as cosmetics and plastics. Most importantly, palm oil will be used to replace diesel fuel, both in motor engines and in industry.

Another 3.7 million acres (1.5 million hectares) are covered with rubber plantations, which used to be the biggest agricultural industry. As land has been converted to growing oil palm and other crops, Thailand has replaced Malaysia as the number one producer of natural rubber. In addition, many industries have switched to using synthetic rubber.

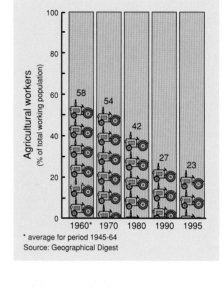

▲ *This graph shows that Malaysia's reliance on agriculture fell between 1970 and 1990, and is still falling.*

◀ *These oil palms stand on the edge of a plantation near Taiping.*

▼ *Palm fruit is squeezed to produce the oil used in many food products, including margarine.*

Farming and fishing

Since the rubber industry began to decline, many farmers with small areas of land have started to grow more cash crops such as cacao, sugarcane, coconuts, and pineapples. Rice forms the staple diet, but people grow it mainly to feed themselves and their families rather than for sale. Fish and seafood are very popular, and many people earn a living on fishing boats along Malaysia's long coasts. However, catches are declining and some fishermen blame this on pollution. To try to protect fish stocks, the government has stopped issuing new licences for fishing boats. Although Malaysia is mainly a Muslim nation and Muslims do not eat pork, it is the biggest producer of pork in the region. Ethnic Chinese farmers run this industry, which survived a terrible outbreak of disease among its pigs in the late 1990s.

▶ *A small pepper tree orchard climbs this hillside. Pepper used to be an important crop in Malaysia, but production is now falling.*

IN THEIR OWN WORDS

I am Abdul Wahab Ismail and I've been growing rubber for almost 25 years. I work at my plantation just outside Taiping. A few years ago, the rubber price was terrible, always going down, but it's a little better these days. Before, we would collect rubber resin from the trees, make it into rubber mats ourselves, then sell it. Now we just sell the resin to middlemen, often processing companies from Singapore. Rubber plantations are being taken over by oil palm. Maybe there's not so much need for real rubber any more, but the future is bright for palm oil industries. I don't know who will run the farms in the future though. Young people don't want to work on farms these days—they want to work in factories or offices, where they earn more money and don't get burned by the sun.

5

The Changing Environment

The cost to nature

Although Malaysia still has large forests, there is concern about the number of trees being cut down. Some people say the timber is being cut down four times as quickly as it would take to grow back again. Many of the accusations of overlogging come from outside the country. The government has responded by saying that countries that have already destroyed their own forests have no right to criticize Malaysia.

Hydroelectric dams, which Malaysia is building to reduce fossil fuel use, are also controversial. In 2001 the government decided to go ahead with construction of the huge Bakun hydroelectric power project in Sarawak. This could destroy even more rain forest. Another threat to the forests comes from fires started by farmers who use the traditional slash-and-burn method of rice growing. This involves burning down large sections of forest to clear the land for planting. Loss of forest threatens wildlife and leads to problems with flooding.

▲ *Huge logs are transported out of the forest. Both natural forests and old rubber tree plantations are making way for construction and development.*

◄ *Here a new dam is under construction. There are environmental benefits as well as problems related to the building of dams.*

IN THEIR OWN WORDS

My name is Apin. I am 22 years old. I live in the village of Kampong Pertak. The village has been rebuilt in a new location because a dam is being built, and the lake would have flooded our old homes. They have cut down the forest to sell the wood before they make the lake. Our old houses were made of wood and bamboo, and we made them ourselves. These houses are made of concrete. Most of us liked the old ones best. We have water in the taps now, but we still like to wash in the river. I sometimes wonder what will happen to me and my village in the future.

Muddied waters

The growth of cities and industries has made it difficult for Malaysia to manage the increase in sewage and waste from factories and farms. Water pollution is affecting many of the country's coastal waters and rivers. Fishermen have complained about property developments that take land along the coast so that it can be built on. This can drastically affect tidal patterns, creating problems for fishermen. In addition, companies that dump industrial waste and mud into the sea are killing fish stocks. The government has introduced various laws to try to tackle the problem, but it seems to have difficulty in protecting fragile ecosystems against business and development.

▲ *Although this water on the coast of Penang looks clear, the increase in tourism to the island is causing the water to become more polluted.*

The growing cities

Kuala Lumpur now has a population of over 1.3 million. The city, often known simply as KL, has stretched out to meet its neighboring towns, forming an urban megalopolis (a city with area around it that is also crowded) called Kelang Valley. Many workers commute from the surrounding state of Selangor, whose population grew by more than 60 percent between 1980 and 1991. Other big growth areas include Sabah and Johor Baharu.

◄ *As Kuala Lumpur grows, high-tech light rail systems are being introduced to try to reduce pollution and traffic jams.*

Fifty-four percent of Malaysians are now thought to live in urban areas, with more moving in from the countryside every year. The growth in population means that services such as transportation, housing, and water systems in Kuala Lumpur and the other popular cities are struggling to cope. The government is trying to reduce the growth of big cities by encouraging people to move to different areas. One of these areas is Putrajaya, 15.5 mi (25 km) south of Kuala Lumpur, which is going to become the new capital. Government offices are being moved there from crowded Kuala Lumpur.

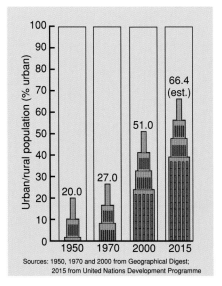

Sources: 1950, 1970 and 2000 from Geographical Digest; 2015 from United Nations Development Programme

▲ *This graph shows that more people are moving to the cities in Malaysia.*

Hazy city

As it grows Kuala Lumpur has become infamous for its traffic jams and air pollution. In recent years the situation has become almost unbearable because haze from forest fires in Borneo and Indonesia has drifted over the city, blocking out sunlight. Much of the capital's pollution comes from thousands of motorcycles, which are popular in Southeast Asia. The cheaper motorbikes have engines that create a lot of noise and smoke. Steps have been taken to improve public transportation and reduce pollution. Three new metro systems are being expanded.

▼ *Traffic pollution and the burning of forests in rural areas cause dense haze in the air over Kuala Lumpur.*

IN THEIR OWN WORDS

My name is Stella Tan Pei Zin. That's me in the middle with my friends, Tan Tiong Shek and Samantha Tan. At school in Kuching, we study recycling in science, but we have only done a few projects. On television they encourage us to collect things and bring them into recycling centers. Some of the materials can be used again, but others have to be thrown away because they may be dangerous to the environment. Plastics take a long time to break down, and you can see bags and other things scattered in rural areas.

Protecting paradise

Although Malaysia is the world's biggest exporter of tropical hardwoods, some measures have been taken to protect its forests for future generations. The law dictates that 69 percent of the remaining forest area is Permanent Forest Estate, with strict rules on how many trees can be cut per acre, and the size of trees that may be cut down. In addition, 12 percent of the forests are now totally protected and form the country's national parks and reserves.

Tourism has become big business in Malaysia, and many visitors come to see the amazing rain forests. The astonishing variety of plants and animals they contain also promises to help humankind's health. Scientists have begun searching the wealth of life in Malaysia's forests for new medicines to combat AIDS, cancer, and other illnesses.

▼ *Hikers in the Borneo jungle wait to catch a glimpse of the elusive orangutan. The word orangutan in Malay means "man of the jungle."*

Balancing act

Protecting natural riches like forests and water life is a huge challenge in a country trying to develop economically. The government wants to make money from wood, but it also wants to reduce deforestation. One way to do this is to encourage businesses to process logs into sawed timber or furniture. These products can be exported for higher prices than cut tree trunks, so fewer trees have to be cut down in order to make money. New plantations, such as those for teak in the north, are being planted especially for use in the timber industry. The plantation trees can be used in the future so that natural forests are not cut for business.

▲ *Destruction of the forests is being carefully monitored by the government.*

IN THEIR OWN WORDS

My name is Anthony Sebastian. I'm an ecologist living in Kuching. Environmental issues are being discussed much more now—people seem to be taking an interest at last. These issues are even included in the school curriculum. Of course this does not always mean that the real issues are being talked about. An achievement in any country is not just big cars, big houses, and money; it is also about having a sense of belonging or ownership.

When people begin to understand that all that surrounds them has to be shared and looked after for everybody's benefit, then the environment will have a chance.

6

The Changing Population

Population explosion

The number of people in Malaysia has doubled since the early 1980s. Although the population continues to grow, its rate of increase has slowed. Improving economic conditions means people have more income and so do not need as many children to help them grow food or earn money. The government's population program educates people on the benefits of planning their families and having fewer children. In 1970 Malaysian families each had an average of 5.2 children. By 2000 that figure had dropped to an average of 3.2 children per family.

One reason for the continued population growth is the country's economic success, which has attracted many immigrant workers, particularly from Indonesia, Bangladesh, and Pakistan. They come to take low-paid or hard jobs that Malaysians do not want.

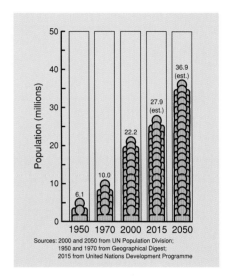

Sources: 2000 and 2050 from UN Population Division; 1950 and 1970 from Geographical Digest; 2015 from United Nations Development Programme

▲ *This graph shows that Malaysia's population has more than doubled in recent years. It is predicted to keep increasing, especially in cities such as Kuala Lumpur.*

◀ *Many families live in high-rise apartment buildings around Kuala Lumpur.*

Longer lives

Life expectancy is also increasing. Malaysia is now a youthful society, but by 2025 it will be an elderly one. In traditional Asian society, families greatly respect the aged and look after them at home. With health care facilities improving though, people are living longer and the country will have to look after many more old people in the future. In 1975 Malaysians lived to an average of 63 years old, but now the national life expectancy is over 72 years. Traditional family life is changing quickly both in the cities and in remote villages. Because modern families have fewer children, there will not be as many young working people to care for the elderly people of tomorrow.

▼ *The family is very important to Malaysian culture. Here the whole family is on an outing at a Kuala Lumpur park.*

IN THEIR OWN WORDS

My name is Payang, of the Bidayuh people. I've lived in Kampung Anna Rais, in Sarawak, for 51 years. Originally we lived in the longhouse with many other families, but I have five children so we built a separate house nearby; many families have done this now. This is a big change in our community. People never used to have separate houses, even though they had more kids before. We are comfortable with our lives. We have always had land and a strong community spirit; we still have a sense of sharing. The youngsters now go out of the community and earn some money, but they almost always come home again. Minority groups like us used to be a little like second-class citizens and we did not venture out of the community, but now we have adapted to the changes in modern society.

Princes and their predecessors

Ethnic Malays, who make up about 50 percent of society, are known as *bumiputera*, or "princes of the soil." Since colonial times much of Malaysia's business has been controlled by ethnic Chinese. After independence this caused resentment among Malays, resulting in race riots. Since then laws have been passed to give Malays more power in economic life. Every business, education, and civil service organization must give a certain percentage of jobs to Malays.

The many ethnic groups who lived in Malaysia long before anyone else are called *orang asli*, the "original people." They now form less than 10 percent of the population. The *orang asli* generally share a strong spiritual tie to the forests. Their lifestyle is changing drastically in some areas because of both forest exploitation and new social services.

▲ *The Bidayuh people of Borneo traditionally live in "longhouses," which can be over 328 feet (100 meters) long and house up to 50 families.*

Economic backbone

The Chinese and Indian communities are often regarded as the country's economic core. As in most of Southeast Asia, the Chinese still have great economic power in Malaysia. Many Indians came in colonial times to work on rubber plantations. Although some have been successful in business, many are still poor. Indian and Chinese temples, architecture, and food, however, contribute greatly to Malaysian culture. In politics the Malays have dominated all groups since independence, but there are now many political parties, seeking to represent everyone.

▲ *A young girl at the Hindu Sri Maha Mariamman temple in Kuala Lumpur.*

IN THEIR OWN WORDS

My name is Wong Karhoo and I live in Kuala Lumpur. I am sitting here outside this Chinese temple waiting for my younger sister to finish her English lesson. Many people take extra classes at temple schools, especially English and mathematics. When I was younger, only Chinese kids went to these schools, but now a whole range of kids from all ethnic groups study there. When I returned to Malaysia after studying in Britain, I was really surprised at the huge changes. Even new cities had been started. There are so many opportunities.

7 Changes at Home

Family life

The people of Malaysia have a variety of lifestyles. Traditional Malay culture centers on the *kampung*, or village, where respect and obedience toward parents and elders is very important. Order and tradition in the village are strengthened through the influence of the community, led by the village headman. As in many other cultures all over the world, though, the traditional ways of life are changing as more and more people move to urban areas. Today many Malays in the cities are adopting different lifestyles. Later marriage and smaller families mean that young people are enjoying greater social freedom and doing fewer home chores.

▼ *Some traditional Malay houses can still be found in the cities.*

Intermarriage between ethnic groups is increasing, and all groups seem to be moving away from the traditionally strict rules of domestic life. The younger generations are less likely to live in the big family groups that their grandparents knew, and in cities it is very difficult for extended families to stay together.

Western influence

Western television shows, movies, music, and advertising have become popular. Many of the people who see them want to adopt the lifestyles presented by the Western media, which seem more modern and exciting. However, there are also many Malaysians who want to preserve their culture. The government sometimes tries to limit foreign media while encouraging traditional arts and ways of life.

▶ *Young people on motor scooters get together during the early hours of a Kuala Lumpur morning.*

IN THEIR OWN WORDS

My name is Faridah Stephens and I live in Kuala Lumpur. In the early 1990s the media situation really changed in Malaysia and many new publications were started. We even got cable TV, which meant we could see overseas programs. It was very exciting, and gave people a wider view of what was happening in the world. The government gives out licences for media publications, and of course, as in most countries in the world, different publications present the views of different political groups.

Newspapers here are printed in the three main languages of the population: Malay, Chinese, and Tamil. The Internet has really transformed publishing and public access to information and news. Now people can read so many points of view and learn so many opinions.

Women's roles

Changes in the economy and society mean that Malaysian women are generally becoming more independent. Among educated people, there is less pressure to get married and have children at a young age, meaning that women have more opportunities to study and work. Previously, women were lucky if they had the chance to finish grade school. Now more girls than boys finish high school, and nearly as many girls as boys go on to college.

▼ *A mix of religions and attitudes can be found among young people in Malaysia.*

Health care for women has also improved. The number of mothers who die during childbirth fell by about 70 percent between 1980 and 1998. Malaysia as a whole still believes in the importance of the family, and in rural areas, where traditions are strong, people marry early. If an unmarried girl gets pregnant, her family will push her to marry.

IN THEIR OWN WORDS

My name is Dr. Ainon Mohd Mokhtar and I live in Kuching. I'm the only female physician in this hospital, but a lot of women are now in the medical profession. I'm not married. This may be partly due to my career and partly because I want to be independent. People accept that the new way is better than before and I don't think I've been pressured by society. Of course a lot of women still put family before their jobs and choose to become housewives, but there are plenty of married women doctors who are able to succeed in both their career and their family lives. Compared to my parents, I should consider myself lucky. During their time it was common for women to have no basic education.

Coming to power

In many Muslim countries, women are not allowed to work outside the home, but in Malaysia women have good access to jobs compared with some of their Southeast Asian neighbors. Women now make up nearly half the work force. Traditional ideas still affect women in some ways—for example, there is a labor law applying to women, stating that they cannot do night shifts or heavy physical work. Over 30 percent of business and administration managers are women, and one of the country's most famous political leaders, Wan Azizah Wan Ismail, is a woman. When she appears in public, Wan Azizah still wears the traditional dress of Malay women, proving that tradition and change can sometimes go hand in hand.

▼ *While the roles of Malaysian women may be changing, many still go to the mosque for daily prayers.*

Education

It is not just women who are benefiting from improved education in Malaysia. The country has made great strides since its independence. The state now provides free education for all children between six and eighteen, and nearly all children attend primary school. Parents can choose to send their children to primary schools that teach in Malay, Chinese, or Tamil, the language of most of the ethnic Indian population. Secondary school students must study in Malay, although in some schools certain subjects are also taught in Chinese or Tamil. After independence English was banned from schools, but that has now changed, and all pupils must take English as a second language.

▲ *Learning English at a language college is seen as essential to getting ahead in business in modern Malaysia.*

Higher education is increasing, with many students choosing to study sciences and technology. Malaysia now has more than seventeen public and private universities, including the National University in Bangi and the University of Technology in Johor Bahru. Some rich families still send their children to foreign universities, but the standard in Malaysia continues to improve.

► *This is a new university campus near Kuching, in Sarawak.*

IN THEIR OWN WORDS

My name is Hafsah Mohdzain. I live and work on Pinang Island. I think I always wanted to be a teacher, but I have to be a student for three years before then. I'm in my final year now and have to do my in-school practice as soon as I graduate. I'm a little worried about it, actually. I feel children have changed a lot from when I was at school—they seem much more aggressive now. Good teachers are so important though. How are kids going to learn if the teachers are not good? We are a racially mixed country and there are different attitudes and ideas, but education and teachers are vital to moving the country forward and helping everybody develop fully.

In rural areas obtaining a good education is more difficult. Language can be a problem for some ethnic minorities, who do not have enough teachers or books in their own language and do not speak Malay well. Finding teachers to work in such places can also be a problem. Some international and government organizations promote what is called non-formal education in these areas. This aims to teach ways of improving living standards to people of all ages in entire communities.

▶ *These young girls attend a rural school run by local religious leaders.*

Diet, health, and medicine

Health care is another area in which Malaysia has made great advances since the 1970s. In the 1990s government policy included a National Health Program that helped to reduce the number of infants who died soon after birth. The program also educates communities and young people about the benefits of family planning and the dangers of unprotected sex.

Modern hospitals have been set up in towns and cities, and there are more doctors than before, although traditional healers (shamans) and traditional medicine still play an important role. Among Malays the shaman is known as a *bomoh*, while the Chinese visit a *sinseh*. Shamans may call on spirits when treating people, or prescribe medicines made from local plants and animal products. Some Chinese medicines are now banned because they come from endangered animals, but Western doctors have started to study some traditional practices.

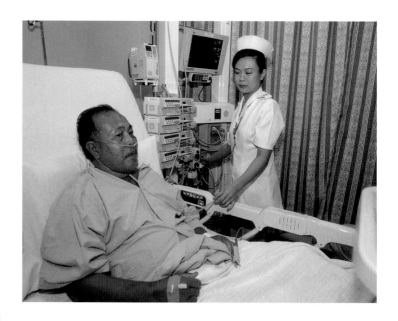

▲ *This modern hospital has state-of-the-art facilities.*

IN THEIR OWN WORDS

My name is Nai Anak Lahai, of the Temuan people of East Malaysia. I'm so old the number is lost now. I'm the oldest survivor of my family. We have always been shamans, working with sick people to make them better. I play a special instrument called a *buloh limbong*, made from bamboo that draws out the evil spirits from a person. Then they get better. Sometimes it takes days and days for this ritual to heal a sick person. I have been playing since I was a small child. It's the only thing I have ever done. These days we don't do it very often. Youngsters still believe we can make them better, but maybe they don't get sick so much these days.

◀ *Fast-food restaurants have become popular in the cities.*

The ethnic mix means that there are several different traditional diets in Malaysia, all of them healthy. The staple foods are rice or cassava, and most protein comes from fish. In cities Western-style food outlets are becoming popular, though many people prefer the local, healthier food. Typically, Malaysians eat a lot of fruit and vegetables, and alcohol consumption is low. Muslim Malays are forbidden to drink alcohol, by both religion and the state, although the law permits non-Muslims to drink.

▶ *A Chinese family enjoys a traditional bowl of tea in Kuala Lumpur.*

Religion

As a multicultural society, Malaysia enjoys a blend of religions, and people are free to follow whatever beliefs they choose. Just over half the population are Muslim, and mosques are a common sight all over the country. Chinese Malaysians tend to be Buddhist or Taoist, while most of the Indian population is Hindu. About 8 percent of the people are Christian. Some minority groups still follow their traditional animist religions, with rituals that are led by village elders or shamans.

▶ *At the Hindu festival of Thaipusam in Kuala Lipis, people smash coconuts in front of a procession to honor Lord Subriaman.*

IN THEIR OWN WORDS

My name is Bhante Aggacitta and I am from Taiping. I've been a Buddhist monk since 1978. I was not satisfied with what the world was offering me. I left Malaysia for thirteen years and then spent a further seven years on a remote mountain with no contact with anybody. When I finally came down I was shocked by what I saw! One person had this small round machine that took shiny discs. I had no idea what it was and was amazed when I found that it could play music. They called it a CD player!

Call to prayer

Malaysia does not interpret Islamic law as strictly as do some Arab and African Muslim countries. Women have a lot of freedom in Malaysia, unlike women in many other Muslim countries. Malays still keep to customs that were in use before Islam was brought to the region, such as consulting the *bomoh*. In some areas the traditional Malay wedding still follows the old Hindu ceremony, and there is generally great tolerance of other religions. Young people across the country appear less interested in religion than the older generations.

Ramadan is one Muslim tradition that affects the entire country. During Ramadan all Muslims fast from sunup to sundown for a whole month. They break their fast with food and drink at sunset. This is a holy month, and even non-Muslims in Malaysia are mindful and respectful of the month of Ramadan.

▼ *Muslim men pray in a mosque on Penang Island.*

8

Changes at Work

Rapid progress

Employment opportunities have become much better for Malaysians as the country's economy has developed. Traditionally people used to have a choice between subsistence farming or working in plantations and mines. People would start work very early in the morning and finish at midday, to escape the heat of the afternoon.

◀ *These days there are thousands of office workers in Malaysia's cities. Modern international-style cafés are popular lunchtime venues for businesspeople.*

Industrial, urban, and social development has dramatically altered the options for young Malaysians. There are jobs in businesses producing raw materials such as natural gas and palm oil, and in manufacturing industries. A wide array of jobs in service industries such as transportation, tourism, and entertainment are available. Most people now go to work at 8 A.M. and leave at 4 P.M. Today only 18 percent of people are employed in agriculture, forestry, and fishing. A third work in industry, and half in services. Unemployment is low, with only 3 percent of the workforce unable to find work in 2003.

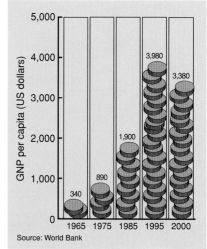

Source: World Bank

▲ *This graph shows that Malaysia's income has risen dramatically over the last 20 years, although the 1997 recession meant that growth slowed down.*

IN THEIR OWN WORDS

My name is Shaz. I studied catering at college before starting work at this restaurant in Kuala Lumpur. The tourist industry and restaurant business have been suffering recently. The world seems to be going through some changes at the moment, and fewer people are traveling because of worries about terrorism. Fortunately, Malaysia attracts tourists even just as a stopover to other places, and it seems safe here. We have a mixture of tourists and locals in here, which is good because it means we don't have to rely on just one type of customer. Local people like to come here to show off a bit. I suppose you could call us a lifestyle restaurant!

Tiger economy

Malaysia is one of the countries that economists call the "tiger" economies. These are the Asian countries that showed amazing rates of financial growth during the 1980s and 1990s. All the tiger economies suffered from a recession called the Asian economic crisis, which began in 1997. Although growth slowed down, Malaysia adapted more quickly than other countries, and its main industries are productive once again. Most of the people who lost their jobs in 1997–1998 are back at work, and people continue to leave behind their small plots of farmland to find jobs in the cities. The economy is growing again and about 10 percent of workers are now employed in the construction of new buildings or facilities.

▼ An old cycle trishaw sits next to its modern replacement, the taxi.

Industrial drive

Since the 1970s Malaysia has made progress in developing its industries. The Malaysian prime minister Mahathir Mohamad led the way, taking personal charge of one of the country's successes, the Proton car. When he announced in 1983 that Malaysia would start making cars, many people did not believe it would be possible. Now, however, the Proton company exports to 50 countries and has bought the British carmaker Lotus.

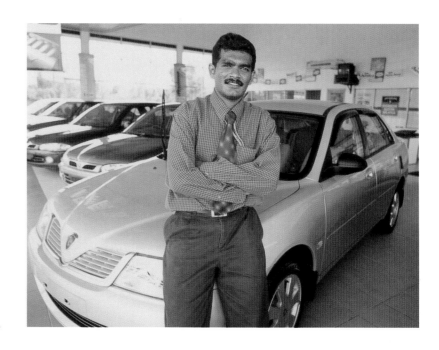

▲ *This man is the manager of a salesroom for Proton, the national carmaker.*

Since the early 1980s, Malaysia has become one of the world's largest producers of disk drives and other electronic components. The country's leaders now want companies to invent and improve, rather than just assemble, electronics. The government is investing billions of dollars to create a light industrial and business zone—the Multimedia Super Corridor—where companies can help Malaysia become a world leader in high-tech industry.

◄ *Cyberjaya, or the Multimedia Super Corridor, has been planned as the business center for the tech industry, but its growth has been slowed by the poor economic situation that affected Asia in the late 1990s.*

IN THEIR OWN WORDS

I'm Mr. Mahadhevan, the supervisor in this palm oil mill in Taiping. Although there has been little change here, other oil mills are much more modern. The palm fruit comes here in big trucks from plantations nearby. We have to cook the fruit and then squeeze the oil out of it—it's a really simple process. The oil is then loaded into tankers and taken away to another factory to be refined. It's a real growth industry with many new palm plantations taking over from rubber trees. This kind of oil is used for many things. A lot of it goes to make margarine, but now they are experimenting on using it for cars.

Thriving mineral industries

In its drive to develop new markets, Malaysia has not abandoned the traditional industries. Tin mines and other mineral concessions are still working. Tin production fell dramatically from the early 1970s to the year 2000. This fall was mainly because the price paid for tin around the world was very low, but more tin will be mined when prices climb again.

Malaysians are proud of their country's achievements, and the government encourages people to support national industries and work for unity.

▶ *The Petronas oil company is very important to Malaysia's economy. It is famous world-wide and even runs a Formula 1 racing team.*

Destination Malaysia

Malaysia is very popular among tourists. The country started to get many more visitors after its Visit Malaysia Year campaign in 1990. Most popular with tourists are Kuala Lumpur and the country's beautiful islands. Langkawi and Tioman islands are famous for their beaches, while Sipadan, a small island off Borneo, is one of the world's most important diving sites. Malacca and Penang attract people interested in historical sites, while ecotourism is a new and expanding industry; people visit the national parks for wildlife walks and caving adventures.

The tourism industry currently has some problems, though. Pollution and development are spoiling some destinations, while fear of terrorism has completely stopped many people from traveling.

▲ *Because of its excellent watersports facilities, the resort of Batu Ferringhi, on Penang Island, attracts many tourists.*

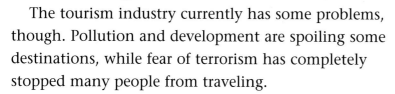

IN THEIR OWN WORDS

I'm Appu, a tour guide in Kuala Lipis. Sometimes we get lots of tourists and then other years just a few—international events seem to affect how many people travel. I often feel like I'm in a one-man fight to develop the ecotourism market. Tourists love the national parks where they can see protected wild animals and amazing plants, but we haven't developed a good system yet. We preserve areas of forest, but nobody has thought about how to market them to tourists. By attracting tourists we can prove that we can earn money from the forests without chopping them down—and this way is sustainable.

Getting around

Malaysia's national airline has a good reputation, and its service has been boosted by the new Kuala Lumpur international airport at Sepang, at the southern end of the Multimedia Super Corridor. The airport has become a regional hub, with people who are traveling long distances using Malaysia as a place where they can change planes, and even stop over for a quick visit.

The country's good road network attracts many visitors from neighboring countries, especially Singapore and Thailand. The railroad system, established by the British, has evolved into one of the finest in Asia and it employs many people. The line carries passengers from Singapore to Bangkok, with the Malaysian section being the most comfortable.

▲ *Kuala Lumpur's new airport has an automatic shuttle train.*

▼ *New infrastructure projects, like this one in Putrajaya, help to improve transportation.*

Shopping heaven

Big shopping centers have sprouted up alongside old-style bazaars, or indoor markets, in Malaysia's towns and cities. International chain stores and restaurants are now common in big malls such as the spectacular Mines Shopping Fair, near Kuala Lumpur. A day trip to a mall is a popular pastime for young people and families, and the malls provide many jobs.

In rural areas people still lead a more traditional life and everyone must share the chores in the house and fields. Most food and household products are bought and sold in markets across the country. Night bazaars, or *pasar malams,* are popular with both local people buying groceries and tourists looking for gifts, handicrafts, or special goods. It is ethnic Chinese people who run most of this kind of business in Malaysia.

▲ *Although Malaysia is a mainly Muslim nation, Christmas shopping is still big business.*

◄ *An old man in a rural village makes a traditional bamboo basket.*

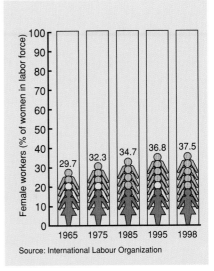

Women at work

Convenience stores and supermarkets have also become a common sight in the cities, and many women work in this sector. Women have always done much of the retail work in Southeast Asia, but the modernization of shopping in Malaysia has meant that such women are learning new skills such as marketing and shop design. Every year more women go to work, many of them in retail shops or restaurants. The number of female shop managers is also increasing.

Female workers (% of women in labor force)

1965	1975	1985	1995	1998
29.7	32.3	34.7	36.8	37.5

Source: International Labour Organization

▲ *This graph shows that nearly 40 percent of Malaysia's working population are women.*

◄ *Both Malaysian business women and female students are experiencing social and economic changes that are widening the areas in which they are able to work.*

IN THEIR OWN WORDS

My name is Harry Bong, and I'm a culture and art specialist from Kuala Lumpur. We seem to have forgotten our own culture and heritage in the rush to be modern and rich. Of course, as the nation develops, things must change, but the recent changes have been so rapid that some people seem to have forgotten to value their own history. It's really about being Malaysian! The past should always give us a reference for the future, and that is why I promote traditional art and modern art in my gallery. Many kids have never been to an art gallery and don't know what our ancestors created.

9

The Way Ahead

The Asian economic crisis of 1997–1998 provided a strong test of the progress Malaysia has made since independence. Many people thought that the government would struggle to keep the country peaceful and stable as the economy declined. However, the economy bounced back after a period of strict controls.

Looking toward the future

Malaysia does have problems to overcome, notably in the areas of energy production and the environment, but the huge investments that the country has made in education and technology should stand it in good stead for the future. Malaysia is a growing economic and cultural presence in the world. For example, a new racetrack at Sepang hosts Formula 1 and motorcycle grand prix. The country takes a strong role regionally as a founding member of ASEAN, the Association

▲ *The giant telecom tower rising above Kuala Lumpur is evidence of Malaysia's modern technological advance.*

◄ *The city of Putrajaya is being built close to Kuala Lumpur. It will eventually become the new capital.*

of South East Asian Nations, and it now gives technological aid to some of its neighbors. Malaysia seems to be on course to keep developing into a prosperous multicultural nation that other Asian countries can look to as an example.

▶ *The education system in Malaysia will provide these students with good opportunities for the future.*

IN THEIR OWN WORDS

I'm Azlan Hajijohan and I'm 18 years old. I am running this Internet and computer repair shop for my brother, who has gone to work in Kuala Lumpur. He taught me about computers, and I learned a lot myself as well, it's not difficult. I want to go and study computer engineering at a special multimedia university in Cyberjaya that is just for studying computers. If you don't know about computers, you really can't do anything now. People in Malaysia have opened their eyes and we can be a center for computer excellence.

Glossary

Aboriginal original or first people to inhabit a country. The term is most commonly used to refer to the original peoples of Australia, but can be used to mean the first people in any area.

animist traditional system of belief based on the idea of spirits that inhabit and protect natural areas, objects, and people

ASEAN (Association of South East Asian Nations) trade association that includes ten countries in Southeast Asia: Malaysia, Thailand, Indonesia, Singapore, Philippines, Brunei, Vietnam, Myanmar (Burma), Lao PDR (Laos), and Cambodia.

Bahasa Malayu main language of Malaysia, which is similar to the Bahasa Indonesia spoken in neighboring Indonesia

biomass fuel fuel made from crops and plants

bomoh Malay word for a shaman (see below)

cacao beanlike seeds used to make cocoa, chocolate, and cocoa butter

cash crops crops that are grown to be sold, often to other countries, instead of being grown for the farmer to eat

cassava tropical plant. Its roots, which contain a lot of starch, can be eaten and used to make tapioca.

deforestation clearing of forests, which are then replaced by other land uses

ecotourism sector of the tourism industry aimed at people who travel to see nature and are often more interested in conservation than beaches or nightlife

federal union system of government in which several states unite into one country or federation while the states still control some laws and policy independently

fossil fuels fuels such as coal and oil, made from the remains of ancient forests and found deep underground or under the ocean

hardwoods heavy and expensive types of wood, such as mahogany and teak, that last for a long time

hydroelectric power electricity generated by turbines that are turned by the force of falling water

infrastructure network of transportation links, communication links, and power supplies that a country needs to have in working order for its economy to function well

logging cutting down the trees in a forest in order to use, or sell, the wood

Malay biggest ethnic group in Malaysia. Malays are nearly all Muslim.

Malaya pre-independence name for Malaysia

mangrove tropical tree that grows in swamps

monsoon seasonal wind that blows in from the oceans, bringing with it heavy rain

peninsula area of land almost surrounded by water, but connected to another larger piece of land

Peninsular Malaysia also called West Malaysia, the part of Malaysia connected to the Asian mainland—the more populated and developed part of the country

recession period when the economy declines, meaning that companies find it difficult to sell their goods for a profit and many people are out of work

resin thick white liquid that oozes out of rubber trees

semiconductor plants factories making devices used in electronic equipment

shaman person in the community who studies the animist spirits and communicates with them at special rituals or ceremonies

slash-and-burn method of farming in which an area of trees and bushes is cut down and the vegetation is burned. The ash that is produced makes the land fertile for crops, however, the land is soon exhausted by the practice.

smog mixture of smoke and fog, or air pollution caused when sunlight reacts with chemicals in vehicle exhausts

subsistence farming growing crops or raising animals to feed oneself, not to sell

sustainable able to last for a long time

trishaw large tricycle with a seat for a driver and a covered two seat passenger compartment

Further Information

Books to read

Department of Geography Staff (editors). *Malaysia in Pictures*. Minneapolis: Lerner Publications, 1997.

McNair, Sylvia. *Malaysia*. Danbury, Conn.: Scholastic, 2002.

Munan, Heidi, and Y Fu. *Malaysia*. Tarrytown, N.Y.: Marshall Cavendish, 2001.

Useful Addresses

Embassy of Malaysia
2401 Massachusetts Avenue, NW
Washington, DC 20008
(202) 328-2700
FAX (202) 483-7661

Permanent Mission of Malaysia
to the United Nations
313 East 43rd Street
New York, NY 10017
(212) 986-6310
Fax: (212) 490-8576
www.un.int/malaysia/

Index

Page numbers in **bold** refer to pages with photographs, maps or statistics panels.